Apollo 8 and Beyond: The Life of Astronaut Frank Borman

Pioneer of Space Exploration and Aviation

Michael D. Maloy

Table of Contents

INTRODUCTION

The Apollo 8 mission stands as a historic moment in the history of space exploration, representing a key step in humanity's voyage beyond Earth. In December 1968, NASA's Apollo 8 spacecraft started on a momentous mission that would permanently reshape our perspective of the cosmos. This mission was not only a forerunner to the epic Apollo 11 moon landing but also a testimonial to the daring and inventiveness of all engaged, particularly Commander Frank Borman.

To comprehend the enormity of Apollo 8, one must rewind to the stormy 1960s. The space competition between the United

States and the Soviet Union was at its pinnacle, with each nation attempting to assert its superiority in the area of space exploration. In this setting, Apollo 8 emerged as a light of hope and success during a difficult era marked by geopolitical tensions and societal upheavals.

The major objective of Apollo 8 was to orbit the moon, a risky endeavor that had never been tried before. This mission was not only a test of technological prowess but also a bold statement of America's resolve to reach the lunar surface. The successful execution of Apollo 8 was vital in preparing the stage for succeeding Apollo missions, culminating in the historic Apollo 11 lunar landing in 1969.

What set Apollo 8 distinctive was its timing and the extraordinary nature of its mission. Launched on December 21, 1968, the spacecraft carried Commander Frank Borman and his two astronauts, James Lovell and William Anders, on a trajectory that would take them to lunar orbit.

The mission not only achieved its scientific aims but also captivated the imagination of people worldwide, delivering a moment of unity and awe during a tumultuous time.

At the forefront of this groundbreaking endeavor was Frank Borman, a name synonymous with bravery, leadership, and a lifelong passion for aviation. Born in Gary, Indiana, Borman's journey from a modest

Midwestern town to the commander's seat of Apollo 8 is a tribute to the American spirit of discovery and adventure.

Borman's early years were defined by a preoccupation with flying. Raised in Tucson, Arizona, he developed a fascination for the skies that would influence his future. His connection with flying began with the assembling of model airplanes alongside his father, establishing the seeds of a passion that would take him beyond Earth's atmosphere.

At the age of 15, Borman took his first solo flight, a remarkable achievement for a youngster in the 1940s. The finances for his flying lessons were hard-earned, as he worked as a bag boy and pumped petrol after school. These early encounters lay the groundwork for a future distinguished by soaring through the skies and exploring the unknown.

Borman's trajectory took him to the U.S. Military Academy at West Point, where he not only got a bachelor of science degree in 1950 but also met his high school sweetheart, Susan Bugbee, whom he would later marry.

Following graduation, Borman went on a prominent career in the U.S. Air Force, serving as a fighter pilot, operational pilot, and instructor. His efforts in the military won him a reputation and laid the ground for his shift to the field of space exploration.

In 1962, Frank Borman was among the elite group of test pilots selected by NASA for the astronaut program. This was the beginning of a new chapter in his life, one that would see him achieve unprecedented heights—both literally and symbolically. The years that followed comprised hard training, spaceflights, and a meeting with a history that would imprint his name in the annals of space travel.

Chapter 1

Early Years: Nurturing Wings of Aspiration

Frank Borman's journey from the sun-soaked vistas of Tucson, Arizona, to the commanding heights of space exploration began in his formative years. The influences of his boyhood, the early seeds of aviation passion, and the rigorous corridors of the U.S. Military Academy at West Point collectively fashioned the character of a man destined for the stars.

Born in Gary, Indiana, Borman's family moved to Tucson, Arizona, where the broad skies and dry landscapes would unintentionally sow the seeds of his future pursuits. Tucson, with its bright skies and wide-open areas, proved to be a playground for young Borman's budding curiosity and adventurous spirit.

The southwestern metropolis, hidden in the Sonoran Desert, provided a distinctive setting for a childhood that would be distinguished by a penchant for travel. Borman's early experiences in Tucson would eventually be reflected in his love for the skies and the great unknown.

Tucson's influence on Borman's perspective was considerable. The beautiful sunsets, the harsh grandeur of the desert, and the proximity to aviation activity in the region led to a sense of amazement that would become a distinguishing aspect of his character.

Aviation Fascination and the Influence of Model Airplanes

Borman's fascination with flying took root in the lush soil of his childhood. The blue skies over Tucson became a canvas on which his dreams of flight began to take shape. One crucial reason for this enthusiasm was the assembling of model airplanes alongside his father.

Assembling these small miracles became a bonding pastime for the Borman family. The precise features of each model, and the careful balance between form and function, inspired a curiosity for the mechanics of flight in young Frank. These early experiences formed the groundwork for a lifelong love affair with aviation, placing him on a trajectory that would travel well beyond the limitations of Earth's atmosphere.

The model airplanes served as both a source of fun and knowledge. Borman's understanding of aerodynamics, propulsion, and the simple exhilaration of soaring to the sky evolved as he methodically stitched together these little flying aircraft.

Little did he know that these early teachings would become the building blocks of a future distinguished by achievements that would transcend gravity.

Flying Lessons and First Solo Flight at 15

Borman's passion for conquering the skies took a real turn when, at the age of 15, he embarked on an adventure that most youngsters could only dream of. With money obtained from after-school work as a bag boy and gas station attendant, Borman invested in flying lessons.

These lessons weren't simply about developing a skill; they were about attaining a desire that had been cultivated since his infancy in Tucson.

Under the direction of expert teachers, Borman learned the subtleties of piloting an airplane. From mastering the controls to learning the mechanics of flight, each instruction pushed him closer to the realization of his childhood goals. The pristine Arizona skies, which had inspired him as a boy, suddenly became the training field for a young pilot anxious to expand his wings.

The climax of Borman's early aviation experience came with his first solo flight. After hours of dual training, the time arrived when he taxied down the runway alone, the weight of the aircraft and the responsibility of flight lying exclusively on his shoulders. The freedom of soaring through the skies solo at such a young age was not simply a testimonial to his skill but also a preview of the endless possibilities that lay ahead.

This early solo flight was an important milestone in Borman's life. It was more than a rite of passage; it was a declaration of independence and a proclamation to the world that he was destined for the heavens. The lessons learned during those early flights would serve him well in the years to come as he rose to greater heights.

Military Career at the U.S. Military Academy at West Point

Borman's trajectory from a solo aviator to a military career took a decisive shift when he enrolled at the U.S. Military Academy at West Point. This renowned institution, with its rich history and tradition of creating leaders, was the perfect crucible for the shaping of Borman's character.

The academy's demanding curriculum, coupled with its emphasis on discipline and leadership, shaped Borman into a strong individual. The ideals imparted at West Point—duty, honor, country—became the bedrock of his military career and thereafter.

As a cadet at West Point, Borman not only succeeded academically but also displayed leadership abilities that would eventually be crucial in his responsibilities as an astronaut and commander. The military training, physical endurance, and mental toughness established at West Point laid the basis for the challenges he would face in the years ahead.

Graduating in 1950 with a bachelor of science degree, Borman had not only gained an education but had endured a transforming experience that would affect his future undertakings.

The discipline engrained at West Point, paired with his passion for aviation, set the stage for the next chapter in his life—one that would see him not only serve his country in the skies but also journey into the undiscovered frontiers of space.

In the crucible of West Point, Frank Borman emerged not simply as a graduate but as a leader, ready to take on the challenges that lay beyond the horizon. His early years had established the framework for a magnificent trip, one that would see him break the boundaries of Earth and explore the cosmos.

Chapter 2

The Gemini Missions: Bridging the Gap to the Moon

The Gemini missions were a critical period in NASA's ambition to conquer space, serving as a bridge between the earliest expeditions into Earth's orbit and the ambitious lunar explorations of the Apollo program.

Frank Borman, a seasoned aviator and astronaut, found himself at the forefront of these missions, contributing to revolutionary feats that would pave the way for humanity's ultimate destination—the moon.

Gemini 7: The First Space Orbital Rendezvous with James Lovell

One of the defining milestones of Frank Borman's trip into space came with the Gemini 7 mission. Launched on December 4, 1965, Borman, along with astronaut James Lovell, began on a mission that would not only test the capabilities of the Gemini spacecraft but also represent a historic first in space exploration—the first space orbital rendezvous.

The purpose of Gemini 7 was to conduct a two-week endurance mission, evaluating the effects of prolonged space travel on the human body. However, what distinguished this mission was its sophisticated dance in orbit with another spacecraft, Gemini 6, flown by Wally Schirra and Thomas Stafford.

The orbital rendezvous was a demanding maneuver that required accuracy and coordination. In a period when space flight was still in its infancy, this feat showcased the increasing capabilities of NASA and the astronauts involved. Borman and Lovell successfully maneuvered Gemini 7 to meet with Gemini 6, marking the first time two manned spacecraft had orbited Earth in close proximity.

The significance of this achievement cannot be emphasized. The success of the orbital rendezvous laid the framework for future space missions, particularly those that would include docking in orbit and ultimately reaching the moon. Borman's role in this historic meeting displayed his piloting talents, agility, and the collaborative spirit that would be important in the years to come.

Challenges and Experiences in the Gemini Program

The Gemini program was not without its hurdles, and Borman, along with his fellow astronauts, encountered a plethora of obstacles as they pushed into the unknown. The Gemini missions attempted to address the problems of long-duration space flight, extravehicular activities (EVAs), and orbital maneuvers—essential components for the approaching lunar missions.

One of the significant problems of the Gemini program was understanding the subtleties of spacewalks. Extravehicular activities allowed astronauts to escape the spacecraft and navigate in the weightlessness of space, a job that necessitated meticulous preparation and execution.

Borman, famed for his calm temperament under pressure, successfully conducted spacewalks during the Gemini missions, leading to the development of tactics that would later be applied during lunar EVAs.

The endurance missions, such as Gemini 7, offered physiological concerns as astronauts spent extended durations in microgravity. Understanding how the human body would adapt to prolonged space travel was crucial for designing the lengthier Apollo missions, including the trek to the moon.

Gemini missions also supplied crucial data on spacecraft rendezvous and docking techniques. As proved in Gemini 7, the ability to rendezvous and dock with another spacecraft in orbit was a vital capability that astronauts would require for lunar missions. Borman's experiences in mastering these approaches further reinforced his reputation as a skilled and adaptive astronaut.

The Gemini program wasn't simply about solving technical problems; it was about pushing the frontiers of human endurance and increasing our understanding of space. Borman's experiences during these flights were not only crucial in advancing space exploration but also in molding his outlook on the unlimited possibilities that lay beyond Earth.

Transition from Gemini to Apollo Missions

With the Gemini missions establishing the capabilities of manned spaceflight, NASA set its sights on an even more bold goal—the moon. The transition from Gemini to Apollo constituted a paradigm shift in the objectives and challenges faced by astronauts, and Frank Borman found himself at the core of this enormous leap.

The Gemini missions had established the necessary basis for orbital maneuvers, spacewalks, and the endurance required for extended space flight. The experiences gathered during these missions were crucial as NASA set its sights on the lofty goal of landing humans on the lunar surface.

Borman's shift from Gemini to Apollo was marked by a strong sense of responsibility. The Apollo missions were not simply about exploration; they were about pushing the frontiers of human achievement and realizing the dream of reaching another celestial body. Borman, along with his fellow astronauts, underwent extensive training and preparation for the challenges that awaited them in the Apollo program.

The change to the Apollo missions brought with it new spacecraft, new technologies, and new aims. The Apollo spacecraft, designed for lunar missions, introduced the lunar module—a separate vehicle that would land on the moon and return to the command module in lunar orbit. Borman and his colleagues had to adjust to this new spacecraft and the difficulties of lunar rendezvous and landing procedures.

As part of the Apollo program, Borman was assigned the job of captain for Apollo 8, the first mission to orbit the moon. This mission represented a huge leap in human space exploration, and Borman's leadership, trained through the hardships of Gemini, would be critical for its success.

The transition from Gemini to Apollo wasn't merely a shift in terminology; it was a leap into the unknown. The Apollo missions represented the culmination of years of study, development, and expertise garnered via the Gemini program.

Borman, drawing on his breadth of experience, became a crucial character in this shift, contributing not just as an astronaut but also as a leader influencing the fate of space exploration.

The Gemini missions were a key chapter in Frank Borman's career and in the greater context of space exploration. These missions tested the mettle of astronauts, pushed the bounds of technical capability, and supplied the crucial stepping stones for the big voyage to the moon.

Borman's experiences in Gemini, from the historic rendezvous in orbit to tackling the challenges of spacewalks, established the framework for the great feats that would follow in the Apollo program.

Chapter 3

Apollo 8 Mission: A Christmas Journey to the Moon

The Apollo 8 mission, a significant milestone in the history of space exploration, unfolded against the backdrop of the space race, geopolitical tensions, and the bold aim of reaching the moon.

Led by Commander Frank Borman, Apollo 8 laid the stage for humanity's first mission beyond Earth's orbit and paved the way for the momentous Apollo 11 moon landing. This narrative analyzes the nuances of the Apollo 8 mission, from its origin to the memorable moments that left an indelible impact on the global mind.

The genesis of the Apollo 8 mission may be traced back to the broader ambitions of the Apollo program. The program, begun by President John F. Kennedy in 1961, aspired to achieve the seemingly impossible—landing humans on the moon and safely returning them to Earth before the end of the decade. As the 1960s progressed, NASA's efforts were concentrated on overcoming the technological obstacles inherent in lunar exploration.

The decision to go for the moon with Apollo 8 was not without its deliberations and hazards. Originally envisioned as an Earth-orbit mission, the success of Apollo 7's mission in October 1968, which demonstrated the durability of the Apollo spacecraft in low Earth orbit, inspired NASA to investigate more ambitious aims. The agency recognized that the time was right to undertake a lunar orbit mission, capitalizing on the momentum gained from earlier accomplishments.

Frank Borman, together with astronauts James Lovell and William Anders, was picked for the monumental undertaking of becoming the first humans to journey to the moon. The decision to deploy Apollo 8 on a lunar orbit mission constituted a deliberate leap into the unknown, driven by a combination of technical readiness, national pride, and the ambition to overtake the Soviet Union in the space race.

Launch from Cape Canaveral and the Journey to Lunar Orbit

On the chilly morning of December 21, 1968, anticipation filled the air at Cape Canaveral, Florida, as the Saturn V rocket carrying Apollo 8 stood poised on the launch platform. The world watched as the countdown hit zero, and the massive rocket roared to life, carrying Borman, Lovell, and Anders toward the moon.

The journey to lunar orbit was a precisely planned route that entailed numerous burns of the spacecraft's propulsion system to attain the precise path required. Apollo 8's path would take it around the moon and back to Earth in a series of carefully coordinated maneuvers. The mission timeline was a delicate ballet of orbital mechanics and human creativity, with the crew reliant on the guiding devices and computations provided by mission control.

As Apollo 8 hurtled across the abyss of space, the crew sensed a strange blend of solitude and connection. The immensity of space, accentuated by the distant lunar destination, highlighted the isolation of their journey. Yet, the connections with mission control, the briefings from Earth, and the camaraderie among the crew members generated a sense of connectedness that transcended the immensity of space.

Navigating over the vast expanse, Apollo 8 approached the moon, a celestial orb that had enthralled humanity for millennia. The lunar module, a separate spacecraft built for lunar landings, remained securely linked to the command module. This mission was not designed for a lunar landing, but rather for a series of circles around the moon before returning to Earth.

The precision of the mission requires split-second calculations and changes. The crew, depending on their training and the knowledge of the mission control team, completed each maneuver with accuracy. The stress and excitement reached a crescendo as Apollo 8 prepared for the critical lunar orbit insertion fire.

On December 24, 1968, Apollo 8 entered lunar orbit, becoming the first manned spacecraft to do so. The achievement was not only a monument to human engineering and daring but also a historic milestone that crossed national boundaries. Borman, Lovell, and Anders were now in orbit around the moon, their spacecraft following elliptical arcs against the lunar backdrop.

Iconic Moments, Including the Christmas Eve Broadcast

The impact of Apollo 8's mission extended beyond its technological achievements; it resonated deeply with people around the world. As the crew orbited the moon, they found themselves in a position to see a vista that no human eyes had seen before—the Earth rising over the lunar horizon.

On Christmas Eve, as Apollo 8 completed its fourth orbit around the moon, the crew sent a live message to Earth. The words stated by Frank Borman during that transmission would resound through the annals of history: "We are now approaching lunar sunrise, and for all the people back on

Earth, the crew of Apollo 8 has a message that we would like to send to you."

Borman, with the Book of Genesis in hand, began reading: "In the beginning, God created the heaven and the earth. And the earth was without form and void, and darkness was upon the face of the deep." In the loneliness of space, the timeless words are connected with the profundity of the moment. The team continued to repeat the biblical scriptures, delivering a message of peace and goodwill to the people of Earth.

The Christmas Eve broadcast from Apollo 8 transcended the realm of science and technology, touching the hearts of people across the globe. In the midst of Cold War tensions, social upheavals, and political struggle, the crew's message of togetherness and optimism provided a temporary relief, a reminder of our shared humanity against the immense cosmic backdrop.

The influence of the broadcast was felt not just by the general audience but also by the astronauts themselves. Borman subsequently remarked on the experience, noting,

"As I looked back at Earth from the lunar orbit, I could see the tension and strife being played out on the surface of the planet we called home. It gave me a profound sense of the interconnectedness of humanity and the fragility of our existence."

"Earthrise" Photo and Its Impact

In the midst of their lunar orbits, the Apollo 8 crew recorded an image that would become one of the most iconic photographs of the 20th century—the "Earthrise." As the spacecraft emerged from the dark side of the moon, Earth appeared on the lunar horizon, bathed in the warm glow of sunshine.

William Anders, the mission's lunar module pilot and photographer, instantly grabbed his Hasselblad camera and caught the amazing scene. The shot represented Earth as a fragile oasis against the bleak lunar environment, a testament to the beauty and frailty of our home planet.

The "Earthrise" shot resonated powerfully with people throughout the world, becoming a symbol of environmental consciousness and the interconnection of all life. The image grabbed the imagination of a generation and fostered a profound appreciation for the Earth as a magnificent and delicate jewel poised in the expanse of space.

The tremendous influence of the "Earthrise" shot extended beyond its aesthetic appeal. It played a significant part in shaping the environmental movement of the 1970s, affecting public perception and policy decisions. The shot, obtained during the Apollo 8 mission, became a strong visual metaphor for the fragility of Earth and the need for responsible stewardship.

In the years that followed, the "Earthrise" shot became an enduring emblem of the Apollo program's greater legacy—a legacy that expanded beyond technological achievements to incorporate a better knowledge of our place in the cosmos and our shared responsibility for the planet we call home.

The Apollo 8 mission, with its daring decision to aim for the moon, its faultless implementation, and the famous scenes that transpired in the lunar orbit, remains a tribute to human ingenuity, fortitude, and the everlasting quest for discovery.

Led by Commander Frank Borman, Apollo 8 not only achieved the technological goals established by NASA but also left an everlasting impact on the collective psyche of humanity.

The choice to send humans on a lunar orbit mission during the stormy 1960s highlighted the tenacity of the human spirit in the face of adversity. The flawless execution of the mission highlighted the brilliance of NASA's engineers, the dedication of the astronauts, and the effectiveness of international teamwork in the pursuit of a common goal.

The Christmas Eve show, with its profound message of unity and optimism, provided a striking reminder of the shared humanity that transcends borders and ideologies. In the quiet of space, the crew of Apollo 8 said words that resonated with people throughout the world, offering a moment of meditation and inspiration amid a tumultuous time in history.

The "Earthrise" shot, with its mesmerizing beauty and symbolic significance, embodied the environmental awakening of the late 20th century. From the desolate surface of the moon, humanity glanced back at Earth, acknowledging the fragility of our planet and the urgency to secure its future.

As Apollo 8 went back to Earth, its crew took with them not only the scientific data obtained from lunar orbit but also a newfound perspective on the interdependence of all life. The mission's influence stretches far beyond the confines of space exploration, affecting environmental conscience, multinational collaboration, and the enduring spirit of human adventure.

In the broad tapestry of space travel, Apollo 8 shines as a light of human achievement, a tribute to our ability to go beyond the familiar and embrace the unfamiliar. The footsteps of Frank Borman, James Lovell, and William Anders resound through the corridors of history, reminding us of the heights to which human aspiration may fly when fuelled by vision, tenacity, and a sense of shared destiny.

Chapter 4

Post-Apollo Career: Frank Borman's Journey with Eastern Airlines

The Apollo 8 mission marked a zenith in the career of Commander Frank Borman, but his adventure did not finish in the expanse of space. After leaving the astronaut corps, Borman stepped into the sphere of commercial aviation, taking on the leadership of Eastern Airlines.

The shift from the thrilling world of space exploration to civilian life offered Frank Borman a fresh set of difficulties and opportunities. The focused and goal-oriented approach that suited him well at NASA would now find application in a different arena—commercial aviation.

In 1970, Borman made a strategic decision to join Eastern Airlines, at that time the fourth-largest airline in the United States. The journey from astronaut to airline boss highlighted Borman's versatility and his desire to embrace new challenges. The choice was not just a career transition; it was a determined move into the vibrant and difficult sector of aviation management.

Borman's selection as the president and CEO of Eastern Airlines positioned him at the helm of a prominent participant in the airline business. The move signaled a departure from the lonely demands of space travel to the complications of managing a large-scale commercial operation. Borman carried with him the leadership characteristics honed in the crucible of space exploration, hoping that they would serve him well in navigating the tumultuous skies of the airline sector.

Leadership Role and Challenges as the President and CEO

Assuming leadership at Eastern Airlines, Borman encountered a multiplicity of obstacles that tested his mettle as an executive. The airline sector, typified by its susceptibility to economic volatility, regulatory difficulties, and operational nuances, requires a leader with vision, strategic acumen, and the capacity to navigate rough skies.

Borman's career as the president and CEO of Eastern Airlines extended from the 1970s until the early '80s, a period marked by substantial changes in the aviation sector. One of his early goals was to undertake cost-cutting measures to improve the financial health of the airline. These actions were vital for Eastern Airlines to remain competitive in a market riddled with problems.

The economic downturn of the 1970s, coupled with the oil crisis and rising fuel prices, provided a tough hurdle for airlines across the globe. Borman, building on his experience in crisis management from his days at NASA, developed strategic strategies to steer Eastern Airlines through these trying circumstances. His leadership was marked by a devotion to efficiency, creativity, and adaptability.

Despite the economic challenges, Borman sought advances in the aviation sector. Under his supervision, Eastern Airlines became the first major carrier to create a shuttle service, connecting key cities in the Northeast. This project aimed at providing frequent, convenient, and reliable air travel for business travelers proved to be a breakthrough concept that set the way for similar services in the future.

However, the path to triumph was not without difficulties. Labor issues, operational challenges, and the expanding competitive landscape increased the complexity of Borman's responsibilities. The delicate mix of handling the requirements of the personnel, preserving operational excellence, and guaranteeing financial viability needed a fine hand at the controls.

Economic and Industry Challenges During Borman's Tenure

The economic issues that marked Borman's tenure at Eastern Airlines were not isolated instances but rather part of a greater industry-wide instability. The 1970s saw the confluence of various elements that greatly impacted the aviation sector, making it a testing ground for executives like Borman.

The oil crisis of the 1970s, sparked by geopolitical developments in the Middle East, led to a dramatic spike in fuel costs. For an industry primarily reliant on aviation fuel, this constituted a difficult financial problem. Eastern Airlines, like its peers, faced with the necessity to absorb rising fuel costs while remaining competitive in a market where ticket prices were under pressure.

In addition to economic factors, the airline sector faced regulatory changes that added a layer of complexity to its operations. Deregulation in the late 1970s brought about a shift in the competitive landscape, fostering new entrants and heightening the rivalry for passengers. Eastern Airlines, under Borman's leadership, had to manage this shifting regulatory climate, modify its business model, and seek new tactics to maintain its market position.

Labor relations also emerged as a serious concern during Borman's term. The airline industry has long been sensitive to labor dynamics, and Eastern Airlines was no exception. Labor conflicts, including strikes and negotiations, brought a degree of complexity to Borman's role as he strove to establish common ground between the interests of the workforce and the imperatives of the business.

These economic and industrial issues required a strategic leader who could make tough decisions while safeguarding the long-term profitability of the airline. Borman's approach, defined by a mix of innovation, cost-cutting measures, and strategic positioning, underlined his commitment to guide Eastern Airlines through the stormy weather of economic instability.

Resignation and Move to Las Cruces, New Mexico

Despite Borman's efforts to shepherd Eastern Airlines through hard times, the economic constraints and operational complexities of the airline sector took their toll. In 1986, after more than a decade at the leadership of Eastern Airlines, Borman took the tough decision to leave from his post as president and CEO.

The decision to retire signified the end of an era for Borman's career in the aviation sector. The challenges he faced, both internally and outside, having pushed the limits of leadership in a dynamic and competitive world. The resignation was not a reflection of personal failure but rather a realization of the developing dynamics of the airline industry and the necessity for new leadership to navigate the changing terrain.

Following his departure from Eastern Airlines, Borman chose a different path, one that would lead him away from the boardrooms of corporate America to the tranquil landscapes of the American Southwest. Borman, together with his wife Susan, went to Las Cruces, New Mexico, seeking a quieter and more introspective chapter in their life.

The move to Las Cruces signified a break from the fast-paced world of aviation and business leadership. Nestled in the gorgeous Mesilla Valley, Las Cruces presented a different kind of horizon for the couple. Borman, ever the explorer, took consolation in the wide-open expanses, the desert scenery, and the quietude of a life away from the spotlight.

In Las Cruces, Borman shifted his attention to other projects. The pair embarked on ranching, creating a cattle ranch in Bighorn, Montana. The transfer from the hectic corridors of airline management to the wide vistas of ranch life signified a major transformation in lifestyle for the Bormans.

The move to Las Cruces allowed Borman to reflect on the many chapters of his life—his formative years in space exploration, his tenure as an aviation executive, and now, his adoption of a more serene living. It was a reflective period, a time to relish the successes, learnings, and challenges that had created his incredible path.

While the decision to resign from Eastern Airlines marked the conclusion of Borman's formal leadership in the aviation industry, it opened the door to a new phase of exploration—this time, an exploration of a different kind, one that celebrated the simplicity of life amidst the vast landscapes of the American Southwest.

Frank Borman's post-Apollo career, notably his stay with Eastern Airlines, serves as a monument to the multidimensional nature of his leadership path. From the commanding heights of space exploration to the subtleties of airline administration, Borman crossed unfamiliar territory with tenacity, adaptation, and a commitment to excellence.

The move from the awe-inspiring missions of NASA to the dynamic problems of the airline business highlighted Borman's versatility as a leader. His strategic perspective, refined in the crucible of space exploration, found application in the complex realm of commercial aviation. As the president and CEO of Eastern Airlines, Borman fought with economic headwinds, industry shifts, and operational intricacies, demonstrating the same courage and commitment that defined his astronaut days.

The economic concerns of the 1970s, coupled with the oil crisis and regulatory changes, presented a formidable backdrop for Borman's leadership at Eastern Airlines. His innovative efforts, cost-cutting measures, and strategic positioning represented a leader who could manage the stormy weather of economic instability while maintaining an eye on the long-term horizon.

However, the airline business, notorious for its volatility and vulnerability to external variables, faced issues that even Borman's leadership could not totally avoid. The choice to quit Eastern Airlines signaled the end of a chapter, leading a move to the serene surroundings of Las Cruces, New Mexico.

In Las Cruces, away from the rush and bustle of business life, Borman found a fresh canvas for investigation. The ranching life in Bighorn, Montana, helped him to reconnect with nature and enjoy a more thoughtful living. The move to Las Cruces signified a departure from the intricacies of airline management to a simpler, more tranquil way of life.

Frank Borman's post-Apollo career, with its numerous chapters and transitions, illustrates the character of a leader who continually pursues new vistas. Whether commanding a spaceship or managing a large airline, Borman's journey displays a dogged pursuit of greatness and a willingness to accept the unknown.

As the final chapter in the book of his professional life unfolded in the landscapes of Las Cruces, Borman's legacy remained inscribed in the sky and the tranquil beauty of the American Southwest—a tribute to the eternal spirit of exploration that distinguishes a real pioneer.

Chapter 5

Personal Life and Reflections: Frank Borman's Journey Beyond the Horizon

The destiny of Frank Borman's life extends far beyond the boundaries of the spacecraft and the boardrooms of airline management.

Behind the famous figure of an astronaut and aviation pioneer lies the human sphere where family plays a key part in crafting the story of one's life. Frank Borman's trip into space, while a testimony to human achievement, also brought forward the intricacies and challenges that space exploration presented on family life.

Borman's wife, Susan Borman, stood as a steady partner through the highs and lows of his career. The demands of space exploration, with its demanding training regimens, extended missions, and the inherent risks linked with space travel, often imposed pressures on family connections. Susan Borman, like many spouses of astronauts, found herself negotiating the uncertainties of her husband's career, bouncing between pleasure in his successes and the worries connected with the inherent perils of space travel.

The Apollo 8 mission, with its historic flight to the moon, marked a zenith in Borman's career but also brought home the reality of family separation. As Borman, along with James Lovell and William Anders, left on the Christmas 1968 mission, Susan Borman and the families of the crew members waited nervously, their hearts interwoven with the fate of the Apollo 8 spacecraft.

The emotional toll of space travel on families is a lesser-known part of the astronaut experience. The solitude of space, the constant prospect of danger, and the responsibilities of a job that exceeds the borders of Earth create a unique dynamic within astronaut families. Susan Borman, with grace and tenacity, exhibited the strength required to handle these hurdles, providing a solid foundation for her husband's efforts.

The impact on family life extends beyond the immediate purpose. The obligations and demands imposed on astronauts, both during training and actual missions, often resulted in protracted periods of separation. The psychological and emotional toll on both the astronauts and their families underlined the sacrifices inherent in the goal of space exploration.

Borman, in his comments on family life during his astronaut years, acknowledged the obstacles experienced by astronaut families. In interviews and writings, he expressed thanks for the constant support of his wife and family, knowing that the pursuit of the extraordinary frequently came at the sacrifice of the mundane moments that make up family life.

Despite the hurdles, the Borman family weathered the complexity of space exploration, finding strength in their friendships and shared devotion to the wider objective of space exploration. The endurance of their family unit highlighted not only the tenacity of the Borman family but also echoed the broader issue of familial sacrifice throughout the astronaut community.

Frank Borman's Perspective on Seeing Earth from Space

One of the most transforming aspects of space travel is the perspective it offers on the fragile beauty of Earth—a perspective that forever transforms the way astronauts regard their home planet. Frank Borman, with his Apollo 8 mission experience, earned a unique vantage position that few have had the honor to witness—the view of Earth from the moon.

The iconic "Earthrise" shot, captured by William Anders during the Apollo 8 mission, depicts the overwhelming impact of viewing Earth from orbit. As the spaceship emerged from the dark side of the moon, Earth appeared on the lunar horizon—a frail oasis poised in the expanse of space. The image caught the imagination of people around the world and became a symbol of the interdependence of all life.

For Borman, the sensation of observing Earth from the moon was a highly personal and emotional occasion. In his comments, he often spoke of the overpowering beauty of the Earthrise—a spectacle that transcended the boundaries of nationalities, ideas, and conflicts. The image, in Borman's words, gave a sense of the shared humanity of humans on Earth, bonded on a tiny blue planet against the backdrop of the cosmic abyss.

The viewpoint gained from space exploration, particularly from the lunar orbit, left an everlasting effect on Borman's worldview. The fragility of Earth, the thin veil of the atmosphere, and the absence of national borders from space underlined the interconnection of the globe. Borman, in interviews and public appearances, underlined the necessity of conserving the Earth as a precious and singular home for humanity.

Borman's perspective on seeing Earth from space goes beyond the scientific and technological dimensions. It delved into the area of philosophy, instilling a sense of responsibility for the planet. The experience of gazing upon Earth from the moon, a perspective only a handful of persons have shared, heightened Borman's understanding of the need for responsible environmental practices and a collective commitment to preserving the delicate balance of life on Earth.

The transforming force of the Earthrise event was not lost on Borman, who regularly remarked about the significant impact it had on his understanding of the human experience. The awareness that, from orbit, the differences and conflicts that define life on Earth are unnoticeable, spurred Borman's campaign for a more harmonious and cooperative global civilization.

Reflections on the Spiritual Aspects of Space Travel

Space travel, with its technological marvels and scientific breakthroughs, also penetrates into the domain of the spiritual—an investigation of the unknown that stimulates thought on the riddles of existence. Frank Borman, throughout his trip through space, encountered moments of transcendence that spurred contemplation on the spiritual qualities woven into the fabric of space travel.

The Christmas Eve broadcast from Apollo 8 remains a poignant example of the spiritual sentiments embedded in space exploration. As Borman, Lovell, and Anders read from the Book of Genesis, the words echoed through the vastness of space, imparting a message of hope, togetherness, and the fundamental interdependence of all life. Borman's part in conveying this message underlined the spiritual undercurrents that flowed through the Apollo 8 mission.

In talks and writings, Borman has commented about the impact of space exploration on his own spiritual perspective. The seclusion of space, the stark grandeur of the lunar terrain, and the quiet that penetrates the vacuum—all these things led to a sense of awe and veneration for the secrets of the cosmos.

The remarkable experiences of space travel encouraged Borman to contemplate the existence of life beyond Earth. The immensity of the universe, with its infinite galaxies and stars, inspired reflection on the possibility of other living forms. Borman's views, while anchored in the scientific curiosity of an astronaut, also touched upon the spiritual implications of contemplating the wonders of the cosmos.

The exploration of space, with its inherent problems and accomplishments, spurred Borman to explore the bigger questions of human existence. The fragility of life, the resilience of the human spirit, and the thirst for knowledge—all these themes became linked with Borman's spiritual views. The immensity of space, as seen from the lunar orbit, became a canvas for contemplating the immense tapestry of creation.

Borman's spiritual reflections stretched beyond the technicalities of space travel to encompass a larger appreciation for the glories of the universe. Whether speaking about the sophisticated architecture of spaceships or the poetic beauty of Earthrise, Borman's words typically expressed a feeling of reverence for the cosmic ballet unfolding in the cosmos.

In the peaceful times of space travel, whether during the transits between Earth and the moon or in the loneliness of lunar orbit, Borman found opportunities for meditation and thought. These spiritual insights, inspired by the unique experiences of an astronaut, offered a nuanced depth to Borman's viewpoint on the human voyage beyond the horizon.

The personal life and reflections of Frank Borman give a rich and multifaceted image of a human whose journey extended well beyond the borders of space. The influence of space exploration on family life, as viewed through the prism of the Borman family's experiences, exposes the sacrifices, hardships, and resilience required to negotiate the undiscovered areas of astronaut life.

Borman's perspective on seeing Earth from orbit, reflected in the classic "Earthrise" photo and his heartfelt observations, speaks to the transforming impact of space exploration. The shift in worldview—from the vantage point of the moon, where Earth is simply a delicate orb against the cosmic backdrop—prompted contemplation on the interconnectivity of humanity and the obligation to maintain our home planet.

The spiritual reflections woven into Borman's journey show the profound and meditative qualities of space travel. From the Christmas Eve transmission to periods of introspection in the quiet of space, Borman's comments touch upon the complexities of existence, the grandeur of the cosmos, and the quest for greater understanding.

Frank Borman's trip beyond the horizon, both in space and in the hallways of personal introspection, stands as a testament to the subtle interplay between the human spirit and the immense unknown. The memory of his experiences relayed through interviews, essays, and the iconic imagery of space travel, provides an enduring source of inspiration for those who continue to stare at the sky and dream of journeys beyond the bounds of Earth.

Chapter 6

Later Years and Legacy: Frank Borman's Journey Beyond the Stars

The later years of Frank Borman's life unfolded as a continuation of his multifaceted journey—one that extended from the commanding heights of space exploration to the serene landscapes of Montana, from the expansive horizons of aviation to the enduring legacy he crafted for future generations of astronauts and aviators.

Establishing a Cattle Ranch in Bighorn, Montana

The transition from the exciting sphere of space exploration and corporate leadership to the quietude of ranch life in Bighorn, Montana, marked a dramatic turn in Frank Borman's latter years. In 1998, Borman started a new enterprise, creating a cattle ranch in the stunning surroundings of Bighorn.

The decision to shift to ranching reflected Borman's fondness for wide-open areas and a yearning for a more contemplative existence. The Bighorn ranch, developed in conjunction with his son, Fred, signified a return to the land—a getaway from the hectic boardrooms of Eastern Airlines to the calm expanses of Montana.

The ranching life, with its rhythms controlled by the seasons and the cycles of nature, gave Borman with a distinct perspective on living. The daily responsibilities of tending to cattle, the expanse of the Montana skies, and the connection to the land became key components of Borman's latter years.

The ranch functioned not only as a financial enterprise but also as a canvas for a new sort of exploration—one that honored the simplicity of living amidst the raw splendor of the American West.

In interviews and publications, Borman acknowledged his enthusiasm for the ranching lifestyle, particularly the sense of fulfillment obtained from working the land and stewarding the cattle. The change to ranch life, far from being a retirement, marked a continuance of Borman's exploratory ethos—an investigation of a new type, grounded in the cycles of nature and the immensity of the Montana countryside.

The Bighorn ranch became a getaway for Borman, a place where he could reconnect with the land and carry on the principles of hard work and stewardship to future generations. The construction of the ranch, in many ways, paralleled Borman's past pursuits—both in space and in business leadership—characterized by a pioneering spirit and a devotion to exploring beyond familiar boundaries.

Continued Passion for Flying and Aviation

While the Bighorn ranch provided a calm setting for Borman's later years, his passion for flying and aviation remained undiminished. Borman, having acquired his pilot's license at the age of 15, continued to take to the skies far into his 90s—an emblem of the enduring love affair with aviation that had characterized his life from an early age.

The fascination of aircraft, with its sense of freedom and discovery, continued to beckon Borman. In interviews, he spoke of the delight obtained from being at the controls of an airplane, the exhilaration of lifting off and soaring through the heavens. The same mentality that carried him through the trials of space research and airline management found expression in the simple pleasure of operating an aircraft.

Borman's continuous commitment to aviation extended beyond personal leisure. His ideas and experiences were sought after, and he became a renowned figure in the aviation community. Whether sharing memories from his early days of flying or delivering thoughts on the progress of aviation technology, Borman remained an interesting and prominent voice inside the field of flight.

The connection to aviation is also reflected in Borman's work with educational programs and organizations dedicated to preserving the legacy of aviation. As a living link to the golden age of flying and the pioneering days of space exploration, Borman's contributions stretched beyond his own aspirations, extending to the greater narrative of human success in the skies.

His continuous passion for flying served as an inspiration for aspiring aviators and astronauts, confirming the notion that the love for flight knows no age or bounds.

Borman's final years, punctuated by flights in both contemporary and vintage aircraft, embodied the ageless fascination of the skies and the irrepressible spirit of discovery that defined his entire life.

Awards and Recognitions, Including the Congressional Space Medal of Honor

The last years of Frank Borman's life were adorned with plaudits and recognitions that emphasized the enormous influence of his achievements on space exploration and aviation. Among the prestigious awards bestowed upon him was the Congressional Space Medal of Honor, a tribute that acknowledged his remarkable service to the

United States and his pioneering role in extending the frontiers of space.

The Congressional Space Medal of Honor, given by President Jimmy Carter in 1978, signified a pinnacle in Borman's series of distinctions. The award, intended to commemorate astronauts who distinguished themselves by their service to the nation in the realm of space, highlighted the gravity of Borman's efforts as a commander of the historic Apollo 8 mission.

Borman's leadership during Apollo 8, his role in setting the trajectory of space travel, and his embodiment of the highest values of astronautics were captured in the Congressional Space Medal of Honor. The award not only praised Borman's technical and leadership achievements but also underlined the symbolic value of his contribution in the broader context of American space aspirations.

In addition to the Congressional Space Medal of Honor, Borman received a multitude of additional accolades and honors throughout his career. These included the NASA Distinguished Service Medal, the NASA Exceptional Service Medal, and the Harmon International Trophy for Aeronautics. Each prize constituted a monument to Borman's continuous commitment to excellence in the domains of space exploration and aviation.

Beyond the medals and trophies, Borman's legacy was inscribed in the hearts and minds of those who gazed to the skies with a feeling of wonder and promise. His influence stretched far beyond the tangible signs of acknowledgment, influencing the mentality of space exploration and motivating generations of astronauts and aviators to push for new heights.

Legacy and Influence on Future Generations of Astronauts and Aviators

As Frank Borman navigated the last years of his life, his legacy became a guiding beacon for future generations of astronauts and aviators. The tenacious spirit, fortitude in the face of adversity, and the pioneering mindset that distinguished Borman's trip left an enduring impact on the narrative of human exploration.

Borman's impact, established in the momentous Apollo 8 mission, extended beyond the technical achievements of space travel. It embodied the intangible qualities of leadership, courage, and a constant pursuit of perfection. His role as the commander of the first voyage to the moon set a standard for future astronauts, emphasizing the potential that awaited those willing to push beyond the bounds of Earth.

In the sphere of aviation, Borman's legacy rang through the corridors of flight schools and airfields. His passion for flying, from the early days of solo flights to the latter years of piloting diverse aircraft, inspired a new generation of aviators. Borman's experiences conveyed through interviews, publications, and public engagements, formed part of the collective story of the aviation community.

Borman's influence also extended to educational efforts aimed at encouraging an interest in science, technology, engineering, and mathematics (STEM). Recognizing the need to foster the next generation of explorers, Borman actively collaborated with schools, museums, and groups dedicated to inspiring young minds to aspire for the stars.

The Borman name became synonymous with the spirit of exploration—an enduring legacy that transcended individual achievements and resonated with the greater human quest for knowledge and discovery.

The ranching life in Bighorn, the continuous passion for flying, the prizes and recognitions—all these components led to a legacy that reached well beyond the personal domain into the collective awareness of people who longed to explore the unknown.

Frank Borman's influence on future generations of astronauts and aviators was not confined to the technical aspects of space and flying. It was a legacy of inspiration, a reminder that the human spirit when propelled by curiosity and dedication, could conquer apparently impossible problems.

Borman's life story, with its chapters on space exploration, airline leadership, ranching, and continuous involvement in aviation, served as a light for those who desired to push the bounds of human achievement.

The last years of Frank Borman's life unfurled like a tapestry woven with the threads of exploration, resilience, and devotion to leaving an indelible mark on the realms of space and aviation. The construction of a cattle ranch in Bighorn, Montana, signified a return to the land—a quieter, more meditative chapter in Borman's journey.

His continued passion for flying and aviation attested to the ageless appeal of the skies and the enduring love affair with flight that defined his entire life. Borman's presence in the aviation world, distinguished by his ideas and experiences, resonated with aspiring aviators and served as a living link to the golden age of flight.

Awards and recognitions, including the Congressional Space Medal of Honor, adorned Borman's career, representing the peak of his achievements in space exploration. These honors underlined not just the technical achievements but also the leadership characteristics and the pioneering attitude that distinguished Borman's role in setting the trajectory of American space operations.

Ultimately, Borman's legacy spread beyond the tangible expressions of recognition. It lived on in the hearts and thoughts of those who looked to the stars with a sense of awe, inspired by the voyage of a man who dared to explore beyond the horizon. The influence on succeeding generations of astronauts and aviators became an intrinsic part of the collective narrative of exploration, a monument to the enduring spirit of a pioneer whose voyage called others to follow in the pursuit of the unknown.

CONCLUSION

As we reflect on the incredible life and contributions of Frank Borman, a star in the fields of space exploration and aviation, it becomes obvious that his journey transcends the bounds of time and space. Borman's legacy is not only a collection of successes but a tribute to the tenacious spirit of human discovery and the enduring significance of a pioneer who dared to seek the heavens.

Frank Borman's long legacy is that of a true pioneer—a visionary whose journey reverberates through the halls of space agencies, aviation organizations, and the collective imagination of those who gaze at the stars with a sense of wonder. His legacy is a tapestry woven with strands of discovery, leadership, and a devotion to pushing the boundaries of human achievement.

At the center of Borman's legacy lies the Apollo 8 mission, a historic adventure that marked the first human journey to the moon. The decision to shoot for the moon, the launch from Cape Canaveral, the 10 orbits around the lunar surface, and the famous moments recorded in the Earthrise shot and the Christmas Eve broadcast—all of these components characterized Borman's legacy as a commander who led his team to new frontiers.

The Earthrise photo, with Earth poised in the expanse of space, became an enduring emblem of the interconnectivity of humanity. Borman's comments on seeing Earth from the moon extended beyond the scientific and technological aspects; they delved into the spiritual and philosophical implications of space exploration.

His part in giving the Christmas Eve broadcast, reading from the Book of Genesis, underlined the human desire for meaning and connection in the cosmos.

Borman's leadership during Apollo 8 set a standard for future astronauts, exhibiting attributes of courage, resilience, and a pioneering spirit. His ability to negotiate problems, both in space and in the boardrooms of Eastern Airlines, displayed a versatility that characterized his legacy as more than an astronaut—it marked him as a leader who could adapt to the intricacies of diverse frontiers.

The construction of a cattle ranch near Bighorn, Montana, was a manifestation of

Borman's connection to the country and a yearning for a more contemplative lifestyle. This stage in his life repeated the spirit of adventure but in a different environment. The ranch became a canvas for a different type of pioneering—an investigation of the rhythms of nature and a commitment to stewardship of the land.

Borman's ongoing passion for flying, far into his 90s, highlighted the ageless fascination of aviation. His impact in the aviation field extended beyond personal endeavors, making him a renowned figure whose insights were sought. Borman's experiences formed part of the common narrative of flight, contributing to the ongoing story of human exploration in the skies.

Awards and recognitions, notably the Congressional Space Medal of Honor, served as external markers of the enormous influence of Borman's accomplishments. These prizes highlighted not only the technical achievements but also the intangible traits that marked his leadership and pioneering ethos. Borman's legacy, as represented in these distinctions, became a source of inspiration for many who followed in his footsteps.

Perhaps the most enduring part of Borman's legacy is the effect he exerted on future generations of astronauts and aviators. His voyage, from the early days of space exploration to the later years of ranching and continual flight, stood as a beacon for others who desired to explore the unknown. Borman's legacy transcends individual successes; it is a communal narrative of human exploration, resilience, and the steadfast spirit of pioneers.

Borman died Tuesday, 7th of November 2023 in Billings, Montana, according to NASA.

Borman also led troubled Eastern Airlines in the 1970s and early '80s after leaving the astronaut corps.

But he was most renowned for his NASA activities. He and his crew, James Lovell and William Anders, were the first Apollo mission to travel to the moon – and to glimpse Earth as a faraway sphere in orbit.

Launched from Florida's Cape Canaveral on Dec. 21, 1968, the Apollo 8 trio spent three days going to the moon, then slid into lunar orbit on Christmas Eve. After they circled 10 times on Dec. 24-25, they headed home on Dec. 27.

On Christmas Eve, the astronauts read from the Book of Genesis in a live transmission from the orbiter: "In the beginning, God created the heaven and the earth. And the earth was without form, and empty; and darkness was over the face of the deep."

Borman closed the transmission with, "And from the crew of Apollo 8, we close with good night, good luck, a Merry Christmas, and God bless all of you — all of you on the good Earth."

Lovell and Borman had already flown together during the two-week Gemini 7 mission, which launched on Dec. 4, 1965 — and, at barely 120 feet apart, completed the first space orbital rendezvous with Gemini 6.

"Gemini was a tough go," Borman told The Associated Press in 1998. "It was smaller than the front seat of a Volkswagen bug. It made Apollo seem like a super-duper, luxury touring bus."

In his autobiography, "Countdown: An Autobiography," Borman said Apollo 8 was initially planned to orbit Earth. The success of Apollo 7's mission in October 1968 to establish system reliability on long-length flights led NASA to believe it was time to take a shot at going to the moon.

But Borman said there was another reason NASA revised the plan: the agency wanted to beat the Russians. Borman said he thought one orbit would be sufficient.

"My main concern in this whole flight was to get there ahead of the Russians and get home. That was a significant achievement in my eyes," Borman revealed at a Chicago presentation in 2017.

It was on the crew's fourth orbit that Anders captured the famed "Earthrise" photo showing a blue and white Earth rising above the drab lunar terrain.

Borman wrote of how the Earth looked from afar: "We were the first humans to glimpse the globe in its glorious wholeness, an intensely emotional experience for each of us. We said nothing to each other, but I was convinced our thoughts were identical - of our family on that spinning globe. And maybe we shared another notion I had, This must be what God sees."

www.ingramcontent.com/pod-product-compliance
Lightning Source LLC
Chambersburg PA
CBHW062334290526
45794CB00005B/2029